To the Moon and Back
A Celestial Coloring Book
Tabitha L Barnett

All images ©2020 Tabitha L Barnett. All rights reserved. No part of this publication is to be reproduced, distributed, or transmiitted in any form or by any means, including photocopying, recording or other electronic or mechanical methods without prior written permission of the artist. Photocopying is permitted for PERSONAL USE ONLY. Absolutely **no commercial use** of any image is permitted.

©2020 Tabitha Barnett

©2020 Tabitha Barnett

©2020 Tabitha Barnett

©2020 Tabitha Barnett

©2020 Tabitha Barnett

©2020 Tabitha Barnett

©2020 Tabitha Barnett

©2020 Tabitha Barnett

©2020 Tabitha Barnett

Live by the sun,

love by the moon.

©2020 Tabitha Barnett

Join the conversation on Facebook:
www.facebook.com/tabbystangledart

If you enjoyed this book, please consider leaving a review on Amazon.

Please post your colored images online using #tabbystangledart or #tabbyb so I can find them easily.

Instagram: @tabbystangledart
Twitter: @tabbyleann
www.patreon.com/tabbyb
www.sellfy.com/tabbyb
www.tabbystangledart.threadless.com
www.redbubble.com/people/tabbyb
www.tinyurl.com/ttavids

COLOR TEST SHEET

COLOR TEST SHEET

COLOR TEST SHEET

CCCC
CCCC
CCCC
CCCC
CCCC

www.ingramcontent.com/pod-product-compliance
Lightning Source LLC
Chambersburg PA
CBHW080504220526
45465CB00006B/2370